David
29.6-78.

Signals for Survival

This book is based on the film 'Signals for Survival' by
Niko Tinbergen and Hugh Falkus, commissioned by the
British Broadcasting Corporation, and awarded
the Italia Prize for documentary films 1969.

Signals for Survival

Niko Tinbergen and Hugh Falkus
Drawings by Eric Ennion

Clarendon Press Oxford 1970

Human beings communicate mainly by speech. No modern human society could function properly without a spoken and written language. And yet even if we were deprived of speech and print we could still communicate with each other, for we can convey a great deal by various gestures, postures, facial expressions, and sounds.

An angry face and clenched fists indicate a mood of aggression and intimidate an opponent. Smiling expresses a friendly attitude and is understood as such. The crying of a baby reveals discomfort and elicits motherly behaviour. This kind of sign language, or signalling, is less elaborate than speech. At the same time, however, it is less parochial, for all members of our species use and understand it, whatever their nationality or even race; whatever the language they speak.

Animals, too, use sign language. Their signals differ from one species to another; but within each species (as in the human species) the language is universally understood. This is a necessity, for the very survival of each species depends on the proper use and understanding of its language.

To illustrate language within a species, we have chosen a very common sea-bird: the lesser black-backed gull.

The lesser black-backed gull, like the herring gull, can be found almost everywhere along the coast. With its snowy-white head and breast and dark slate-grey wings, it is a bird of rare grace and beauty. But also (like the herring gull) it is a bird of very strange antics.

These curious antics are not accidental. On the contrary, they are language: a system of signalling, comprising posture, movement, sound and colour. When we begin to understand this language the gulls' behaviour assumes a tremendous new interest, and we enter a fascinating world—the world of animal communication.

This is the southern tip of Walney Island off the Lancashire coast; a nature reserve within sight of Barrow's shipyards. Standing here on a winter's day, we see little of special interest. The great sweep of sand and grassland is deserted. There is no sound except for the distant sea.

But in early spring, Walney erupts into a scene of frantic activity. From mid-March until late July the dunes are dotted white from end to end with over 40 000 nesting lesser black-backs and herring gulls. The reserve is dominated by the gulls. Nevertheless, their breeding ground is shared with such colourful shore-birds as oystercatchers, ringed plovers, eiders, and shelducks.

If we walk openly through this vast colony, we see nothing but a cloud of screaming, frightened gulls. The scene appears to be one of utter chaos.

But if we have patience, and take the trouble to watch without being seen, we soon find that there is nothing chaotic in a gull colony.

After a helper has put us in our hide and been seen to walk away, the birds seem to forget that we have stayed behind. Very soon, they settle down again. Then, by cautious observation through the screen of marram grass covering the hide windows, we can see how they behave when undisturbed.

At first we are completely baffled by the seemingly senseless goings-on. But after a while we begin to realize that this great concourse of gulls is no haphazard gathering. It is, indeed, a bird 'city'; a stable social structure, extremely well organized by its inhabitants. Each posture, each call, is part of an intricate language by which birds communicate with one another and make their intentions understood. By doing so, they achieve the co-operation that is vital for successful reproduction.

But a gullery is no city of friends. It is, indeed a city of thieves and murderers. When an approaching predator forces parent birds to leave their eggs unguarded, a bold neighbour may rush in and help himself.

The gulls will even attack and eat their neighbours' chicks. A chick is easily killed by a few pecks from the strong bill—and soon disappears down the astonishingly wide gullet.

How are the effects of these predatory inclinations reconciled with the need to reproduce successfully, and with living in colonies? It seems incredible that such cannibals can raise their broods when packed so tightly together, their nests often no more than a few yards apart.

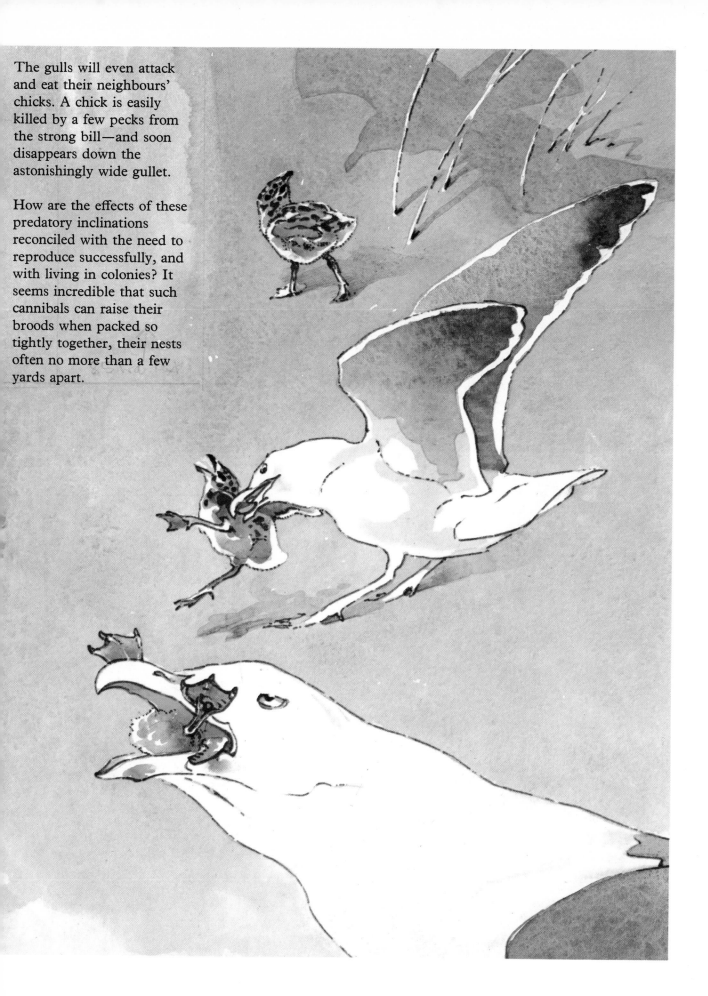

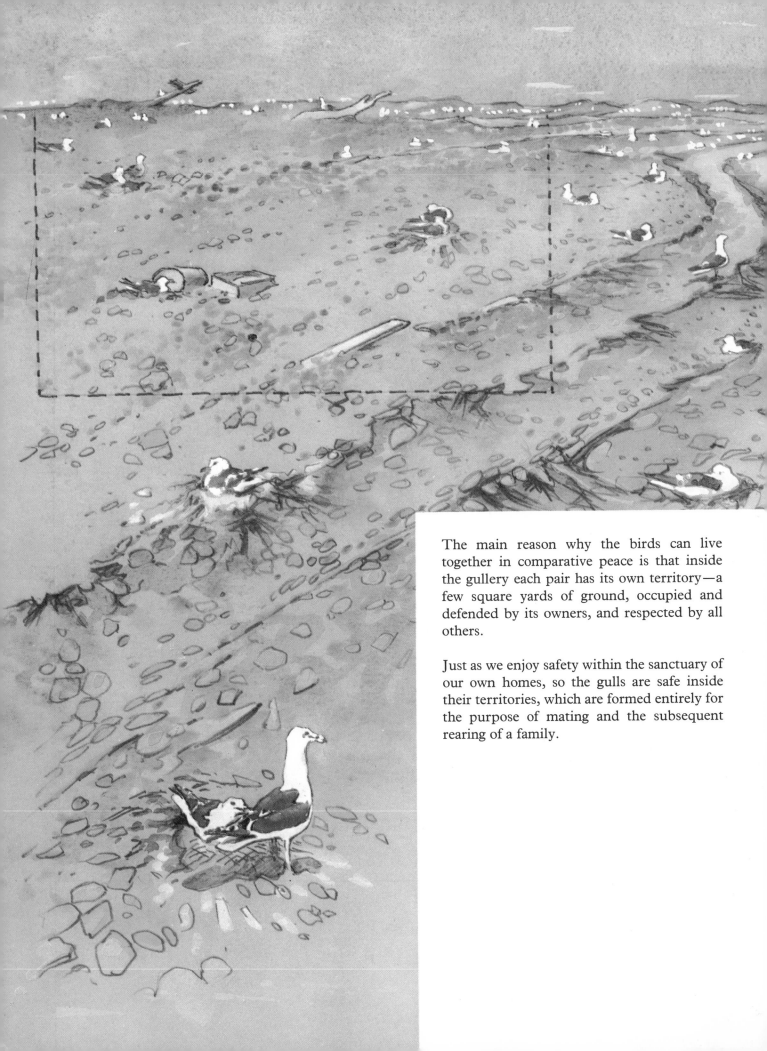

The main reason why the birds can live together in comparative peace is that inside the gullery each pair has its own territory—a few square yards of ground, occupied and defended by its owners, and respected by all others.

Just as we enjoy safety within the sanctuary of our own homes, so the gulls are safe inside their territories, which are formed entirely for the purpose of mating and the subsequent rearing of a family.

The territorial boundaries, so real to the birds, are invisible to us . . .

. . . But if we *could* see them, they would look rather like this.

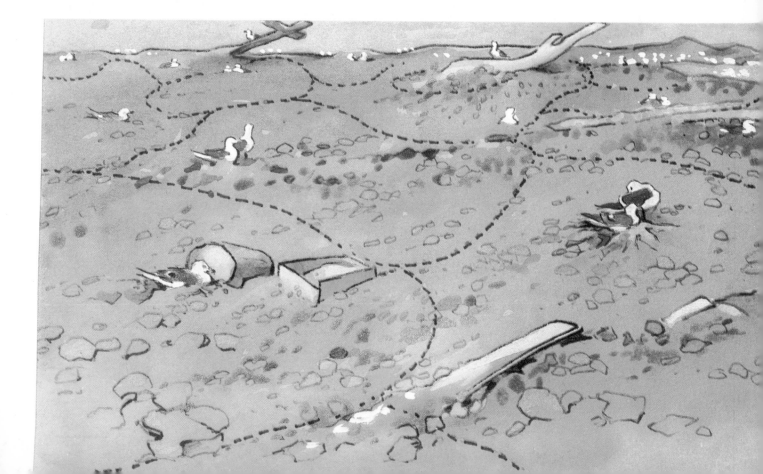

The owner of a territory is always the male bird. Early in the breeding season he stakes his claim to a piece of ground by strutting about aggressively . . .

. . . and by long-calling: a loud trumpeting call, which means: 'This is *my* ground. Keep off!'

PHOTOGRAPH LAWRENCE C. SHAFFER

While territories are being contested the gullery is full of sound. As soon as one male long-calls, others join in, each signalling to neighbours and newcomers alike: 'This ground is occupied!'

Would-be settlers understand this chorus of signals and move on. Neighbours learn which areas to avoid.

C

Usually the signal of ownership is sufficient to discourage trespassers. Sometimes, however, it is not, and the birds will come to blows. A male is prepared to fight any other male in defence of his territory.

A clash seldom goes beyond a 'pulling' fight: a trial of strength, rather like a tug-of-war. Two neighbours will reach across their common boundary and seize each other by the bill, each trying to pull harder than the other. The weaker bird is cowed by being dragged out of his ground into a less secure situation.

Occasionally, fighting can be more vicious. When two males contest the same piece of ground, neither is prepared to give way. Such fights, usually very well balanced, may continue for many minutes, the birds attacking each other time and again.

When a fight is in progress, neighbouring gulls gather round to watch. This too is part of territoriality. The onlookers are ready to interfere if, in the heat of battle, the two contestants blunder into their territories. All birds are continuously on 'border guard'.

PHOTOGRAPH LAWRENCE C. SHAFFER

But serious fighting is rare. This is because the division of breeding space does not depend only on the aggressiveness—however striking—of the territory owners. Fear, too, plays a part, and a big part, in territory establishment. A male still in search of a territory is afraid of all males already in possession of one. If a prospecting bird lands on another's territory, he will usually run away at the slightest sign of hostility.

Such 'cowardice' in non-occupiers is just as useful as the aggressiveness of owners, for it forces each repelled bird to move off and try his luck elsewhere. It is therefore the intricate interplay of aggression and timidity that ensures the quickest possible occupation of vacant sites. In addition it reduces bloodshed; the procedure is economic for both sides.

The well-adjusted interplay between aggression and fear remains evident even after territories have been established, for even well-established males are afraid of crossing their boundaries and moving into neighbouring territories.

Of course, whenever a male approaches his neighbour's territory he is, to that neighbour, a 'near-intruder', and as such is viewed with extreme hostility. Time and again birds meet face to face over their common boundary, each itching to attack the other, yet frightened to do so. Each of them is in a state of conflict between aggression and withdrawal. We see this conflict expressed in the very postures they assume. Fear keeps both birds on their own ground, but both have their weapons of attack 'half drawn'.

A gull's weapons are the bill and the wings. When fighting, a gull stretches his neck and tries to peck at his opponent from above. If he succeeds in holding his opponent down, he half-folds a wing and, using it as a club, delivers violent blows with it. And it is precisely these weapons, poised in readiness, that are conspicuous in the gull's hostile postures: the wings pushed forward; the neck stretched upward with the bill pointing down.

It is remarkable that these postures not only express a bird's hostile intentions but are understood by all the other gulls. They act as signals; as deterrents. And through these 'threat' signals, the attack–escape system prevents bloodshed even between neighbouring well-established males. The final result is an orderly spacing of homes.

Like human city-dwellers, the gulls are drawn together by their sociality, but a mutually understood code of signals expressing ownership keeps them well apart.

This is the 'upright threat posture', used to warn off incidental intruders. The neck is stretched up, with the bill pointing downwards ready to strike. The wrists of the wings are pushed aggressively forward ready to beat with. The eyes are half-closed. This is probably for protection—but how intimidating it looks!

This threat forms a very important part of the gull's language. It means: 'Get out—or else . . . !'

Compare the upright threat posture with the attitude of a gull at rest. Here the bird is quite relaxed; the neck withdrawn, the wings tucked away in the supporting feathers of the flanks.

He is not asleep, however, and any outsider approaching his territorial boundary is instantly detected.

At once, the neck shoots up and the eyes open fully. He is now in the 'alert' attitude, while he assesses the intentions of a stranger. It is only when he prepares to meet an intruder that the 'eye-brows' are lowered and the wrists of the wings are pushed forward.

This male has just charged an intruder and driven it off. With his long wings impressively spread, he rubs in the message: 'No trespassing—keep out!'

Each male signals this warning time and time again. A territory has to be continuously guarded. As the season advances, newcomers in search of vacant sites keep trying to push in. Repelled by male after male, they establish territories either on vacant plots or on the colony fringe.

Early in the breeding season, hostile clashes are taking place all over the colony as males already in possession of territories warn off would-be intruders, or defend their boundaries against their neighbours. Longing to fight, but inhibited by fear, they posture in attitudes of threat. But the furious owners simply have to let off steam; their pent-up aggression *must* have an outlet.

And so, afraid of his opponent but compelled to have a peck at something, each bird pecks violently into the ground and pulls out beakfuls of grass—a movement very similar to pecking his opponent, but aimed just out of reach of his opponent's bill. In other words, he makes a 're-directed attack'—as we, when angry, may bang the table or kick a chair.

A 'pulling' fight. An action similar to grass-pulling.

Grass-pulling in front of an opponent acts as a more emphatic type of threat. It conveys the message: 'Not a step nearer—or else . . .!'

By means of calls, threats, and occasional fights, each male secures a breeding territory in which a future family can grow up un-molested by its neighbours. His next step is to attract a female. As a land-owner, he is now in a position to do this.

Since gulls pair for life, many males already have partners. But there are males whose partners have been killed during the winter. There are young birds, both males and females, that are going to breed for the first time. Each of these birds tries to find a mate.

Females take the more active part in pair-formation. During the early weeks of the season they fly to and fro over the colony on the lookout for suitable males.

Males stay on the ground guarding their territories. They advertise themselves by long-calling—pointing their great trumpet-like throats at an approaching female so as to project the greatest possible volume of sound in her direction.

A male attracts a female by the very same behaviour that repels other males: by being masculine and aggressive, and by broadcasting the signals that proclaim his territorial ownership.

A female glides low over the colony. Below, on his territory, a male catches sight of her and calls.

She lands near by; fairly close to him, but not too close. She is far from certain of the reception she is going to get.

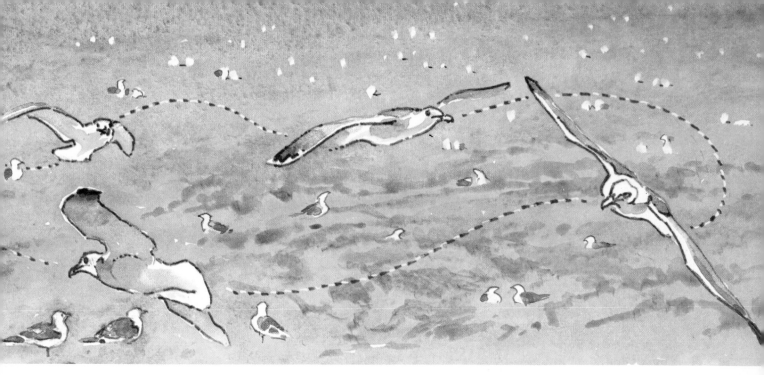

The male's response is mixed. He is torn between desire and aggression. She is a female and therefore desirable. But she is also a stranger to intrude into his territory.

He may accept her at once, or he may peck her away. In defence of his territory, a male has been in an almost constant state of tension. Now, any gull that approaches his jealously-guarded boundary—even a female—is in danger of being attacked.

The female tries to overcome the male's aggressiveness by performing an elaborate ritual of 'appeasement'. Adopting a curious hunched attitude—body and neck held horizontal, head withdrawn between the shoulders—she walks in circles round and round the male, making little tossing movements of the head and uttering a soft 'begging' call each time she does so.

The 'hunched' posture, which signifies: 'I come in peace.' How different from the 'upright threat' posture (p. 23).

'What sort of mood is he in?' A female keeps an anxious eye on her future mate.

Still keeping her distance, the female [on the left] continues to walk round and round the male.

Every now and then she will suddenly stop and turn her head away from him. Like the hunched posture and the head-tossing, this 'facing-away' from the male is the opposite of threat. The bill is a gull's major weapon of attack. Instead of pointing it at him, she hides it. Thus, facing away means: 'I am friendly'.

Gradually the female becomes more confident. Still uttering the begging call she edges closer and closer to the male.

At last she goes very close and, greatly daring, starts to peck at the base of his bill, tossing her head up between pecks. This is the language of love. In a sense, it is 'cupboard' love—since what she is asking for is a meal. Head-tossing combined with bill-pecking means: 'I want to be fed'.

Soon, a swelling appears in the male's neck.
As this swelling grows larger, the male [on the
left] tends to turn away from the female; but
she scurries round in front of him, keeping as
close as possible in greedy anticipation.

Then with what looks like a tremendous effort he regurgitates a huge mouthful of half-digested food—which she quickly gobbles up.

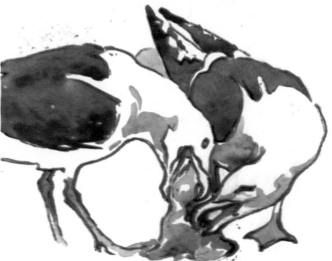

By feeding the female, the male has shown his acceptance of her as a prospective mate. He may still peck at her now and then, even chase her away, but she will return. Once a pair has reached this stage, they are, so to speak, 'engaged'.

But the bond between the two birds takes time to cement. The male never relaxes his defence of the territory. This takes priority even over love. If an intruder approaches, courtship is immediately interrupted.

Here, the male is having an angry argument over the territorial boundary with his hated neighbour. Not very tactfully, the female has followed him, begging hopefully for a meal.

To be pestered at such a moment by a food-begging female makes the male angrier still. Furious with his neighbour—whom he dare not attack—and longing to have a peck at someone, he turns on his future partner. Grabbing her by the neck, or wing tip, he swings her round and round. She makes no attempt to fight back, but struggles to get away, squawking in distress.

Such 'whirligig' scuffles occur even among the flocks of gulls on the low-tide beach. Circular sets of tracks like these are quite common. They leave no doubt as to what has happened.

Early in the breeding season, the female is always in danger of being attacked by the male. This danger must be overcome if the birds are to mate successfully. In this, the courtship feeding of the female by the male, with all its preliminaries, plays a very important part. Feeding takes place two or three times a day, and continues for several weeks. The intimacy and frequency of these meals gradually strengthens the pair bond, and all this extra food undoubtedly helps the female to grow three large eggs in a very short time.

Almost invariably, a session of courtship feeding is followed by copulation. This demands additional language. Having fed the female, the male himself will begin to make the head-tossing 'begging' signals. If the female is willing to mate, she too will head-toss, both birds uttering the melodious begging call.

This signalling is essential because, unlike male mammals, a male gull cannot force a copulation. For copulation to succeed, the female must be willing to co-operate.

Very rarely, a male may try to force a copulation. When this happens, the female pecks him away, preventing him from making cloacal contact. This seldom happens between the members of a pair. The male shown below is trying to rape a neighbour's female who is already sitting on eggs.

Once mated, the female has to learn the boundaries of the male's territory. She does this by watching his disputes with his neighbours. Here, two males are engaged in a 'pulling' fight, watched by their mates. While learning, a female watches a dispute but takes no active part in it. Soon, however, she will attack the opposing female.

What has hitherto been a territory for one has become a territory for two. Long-calling in unison, male and female challenge the outside world together.

Ever since the start of courtship, the male has been taking his mate 'house-hunting'. Uttering a mew call—the 'come hither' signal; a wailing, long-drawn cry which is remarkably like a cat's mewing—he leads the way to a suitable nest site.

At the selected place he bends forward and starts to bob his head up and down, uttering a muffled, rhythmic sound. This signal is called 'choking'—so named because it gives a listener the impression that the bird is short of breath.

His mate joins him, also choking over the same spot. This mutual choking is an important signal. In human terms, they are telling each other: 'This is the place to build the nest'.

The birds 'choke' together time after time, and so come to agree on the exact location of the nest. This is essential, because soon they will have to take turns to incubate and guard their eggs. Repeated choking helps to imprint the memory of where the eggs are going to be laid.

After an argument with their neighbours, a pair will often run, mewing, to the nest-site and have a good choke. This serves a double purpose. It tells their neighbours to 'keep off!' and helps to cement their own friendship.

Their boundaries have been threatened. They have fought the enemy in a common cause. Now they are at peace, together.

By now the two birds are almost inseparable. Together they fly to the feeding grounds; together they feed; together they return. With the days lengthening, there is plenty of spare time. Much of it is spent dozing, either on the territory or on the sea-shore.

In the mornings, particularly after rain (when the ground is damp and nest material flexible), the birds will build their nest. This is usually quite a simple affair: a scrape in the ground with straws woven round the rim.

Often, a nest will lie beside a piece of drift-wood, or even an old tree stump washed up by the sea.

To some birds, tangles of twine and bits of netting from the tide-line seems irresistible—although good use is not always made of this strange nest material.

Once their first egg is laid the birds must conform to a new rhythym of life. They incubate in turn, and can no longer fly out to the feeding grounds together. To ensure that the nest is never left unguarded the sitting bird waits until its partner returns, but the returning bird must signal its willingness to take over. It does this by walking up to the sitting bird with a beakful of nest material; or by mewing; or by choking; or by head-tossing. These are the signals of nest-relief.

Usually the sitting bird will get up at once and allow the returning bird to take over. But the urge to incubate is very strong, and occasionally the relieving bird has to push its partner off the nest.

If a bird leaves the nest without waiting for its mate to return, a neighbour or a marauding stranger will soon carry off and eat the unprotected eggs.

At this stage of the breeding season, when most gulls are incubating, it is striking how peaceful the gullery has become. The raucous sound and fury of the early days, when territories were being so hotly contested, has dwindled to a quiet background chatter. Those strident calls of challenging males no longer drown the skylarks, singing high overhead in the May morning sky.

The rich vegetation breeds many insects, which in turn are preyed on by visiting jackdaws that fly in from their own breeding sites. One eye ever on the nesting gulls, the crafty jackdaws pick their way gingerly through the very heart of the colony.

The free-roaming cattle cannot be expected to know the gulls' boundaries. The gulls accept them as local 'furniture', threatening them only when they graze too close to a nest. Although the cattle do not understand gull language, they soon learn how to avoid a sharp peck on the nose.

After four weeks' incubation, the eggs begin
to hatch. Already the chicks begin to give faint
squeaking calls—to which the sitting parent
reacts. The chicks, too, can hear their parents'
voices. They stop squeaking when the parents
call the alarm.

Cracks appear in the shell, made by the chick's
'egg-tooth', a sharp knob on the tip of the
upper mandible. The chick moves slowly
round inside the egg and so makes a series of
cracks.

After a day or two an opening appears. One
can see the chick's bill-tip through it.

Finally, the chick begins to make rhythmic stretching movements which lift off the 'lid' at the obtuse end. Struggling with its wings and feet, the chick works itself out of the egg.

A newly-hatched chick is wet, but after a few hours' brooding the plumage is dry and wonderfully fluffy.

All over the southern end of Walney Island, eggs are pipping and the downy young of many species tumble out into a chilly world. Young oystercatchers hatch in a less helpless stage than gulls. They can stand very soon after hatching, and are firmly planted on their sturdy legs even while still drying under the adult.

A few hours after they have hatched, young oystercatchers walk off on to the open shore with their parents—who feed them as they go along.

Eider ducklings, too, move off soon after hatching. They follow their mothers, who, often helped by females that have lost their broods or failed to lay, manage to defend their young against attack by the predatory gulls. The ducklings can swim expertly as soon as they reach the water, and begin to feed independently, drifting in and out with the tide.

Eider ducks are remarkably successful in defending their broods against the gulls, and so can co-exist with them.

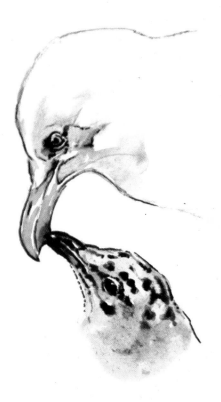

With the hatching of the young gulls, new signals come into action. The chicks have to be fed. But this is not quite so simple as it sounds. The chicks must know where to look for food, and the parents must be told when the chicks are hungry.

To get a meal, a hungry chick pecks at the parent's bill-tip. This pecking makes the parent feed the chick.

The chick does not peck at random. It aims at the red spot on the parent's lower mandible. This red spot, therefore, is a signal; but this time a colour signal, similar in a way to our own flags and traffic lights. In response to this colour signal, the hungry chick signals to the parent by pecking. The parent, in turn, responds by regurgitating food—a 'conversation' of reciprocal signalling.

We need to learn our colour signals; but a chick need not learn to peck at the red spot on its parent's bill. It *knows* where to peck without ever having seen another gull. This 'ready-made' response to the colour red prevents chicks from wasting their energies by pecking at any near-by objects; for instance, at the marram grass, or other vegetation surrounding the nest sites.

But when faced with some of the huge meals offered by the parents, even the hungriest chick can become gorged. This chick literally bulges.

The parent mews, trying to stimulate the chicks to ask for more food. But since they cannot eat another mouthful, they ignore the invitation and refuse to make the appropriate signal.

Usually the parent stops pressing them, and will not give up more food until they once again signal their hunger by pecking at the bill-tip.

The parent birds still take it in turn to guard the nest-site. The chicks must be kept safe, not only from outside predators but from their grown-up neighbours. Although for a time they seldom stray very far, they soon get the urge to explore.

But to venture outside their territory is fatal. Safety lies at home. There is death everywhere else. Any chick that strays too far is invariably killed—like this one, which wandered from the nest above.

But although predatory neighbours are [a]
threat, sociality has its advantages. Many pair[s]
of eyes see more than one pair. When pre[-]
dators appear—and man is one of them—
birds warn each other by calling the alarm:
staccato 'kak-kak-kak', which means 'Loo[k]
out!'

Like life in human cities at the sounding of a[n]
air-raid warning, life in 'Bird City' is suddenl[y]
disrupted. At the approach of danger, a rippl[e]
of alarm spreads through the colony and floc[k]
after flock of parent gulls take to the air[.]
Wherever a predator goes, it is surrounded b[y]
a cloud of screaming gulls.

The young birds, too, understand the alarm signal. Chicks barely able to walk rely on their cryptic colouring and crouch where they are, unmoving.

Older chicks take cover in the vegetation, which has grown high around many of the nest-sites. During a colony disturbance, any movement can be fatal. Predators have sharp eyes. So have the hundreds of adult gulls swooping overhead—among whom only a chick's parents are its friends.

This chick did not hide promptly enough, and paid with its life.

When danger has passed by, the parent gull alights and calls the chicks by mewing. Just as they have to be told when to crouch, so they must be told: 'All is safe'.

Each chick recognizes its parent's voice and responds to the 'come hither' signal. Soon they are all back in the safety of the nest-site, and family life is resumed.

The young gulls grow very rapidly, and their food requirements become enormous. Their constant pestering for more and more food urges the parent gulls to a tremendous effort.

At the first hint of dawn, the adults start their long working day as they stream out in ones and twos towards the feeding grounds. By the time the sun appears over the horizon, hundreds of fully-loaded gulls are already flocking back.

For hour after hour throughout each long summer's day the gulls drift in and out of the colony, seemingly unhurried and casual, and yet successful in ferrying a huge supply of food with unfailing regularity.

By the time a young gull is half-grown, its food-begging signal has changed. Now it pecks at the base of the parent's bill—instead of pecking at the bill-tip, as it did when it was a chick.

At the peak of the breeding season roughly twenty tons of food a day are being flown into the gullery, whose numbers have now swollen to some 80 000 adults and young. Twenty tons a day seems an astonishing amount. But along the sea-shore the gulls find plenty of pickings—such as this stranded lumpsucker.

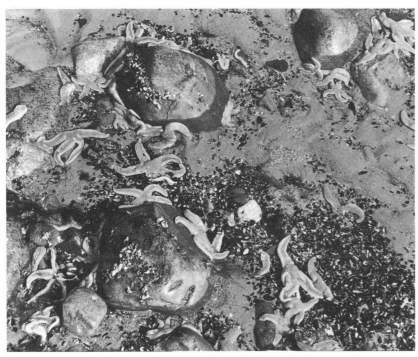

Starfish and small herrings left behind by the tide.

Some gulls fly to inland fields for worms or carrion; others to fish docks and refuse-dumps for offal. But since the lesser black-backed gull is primarily a fisherman, many birds fly out to sea to follow the fishing boats, or to fish for themselves.

The food remains left strewn around the nest-sites tell a story of the gulls' omnivorous habits. Some birds, however, are specialist feeders and prefer certain types of food.

As we can see from these bones and vertebrae, this bird is a fish-eater.

And this one, a crab specialist. Dozens of tiny shore crabs have been regurgitated for the young, but not eaten. PHOTOGRAPH LAWRENCE C. SHAFFER

Heaps of tiny shellfish (baltic tellins), which the gull has found on the mud flats.

Chicken's feet from the local refuse-dump.

Some habitués of the refuse-dump swallow butter papers. These are regurgitated as pellets, and thousands of wrappings end up in the colony. Time and again one sees the young try to eat, and then reject, these useless presents.

This nest-owner, another regular at the refuse-dump, has specialized in swallowing milk-bottle tops. By the time it reached its nest, the bird would have digested the cream, and the useless metal—which may have been regurgitated as a meal offering—must have been greeted with dismay by the ravenous young.

While guarding the young, or resting on the beach, the birds spend hours preening their plumage. During the breeding season each gull has a tremendous flying programme to carry out. Like that of any aircraft, flying machinery must be kept in perfect order.

Feathers must be waterproofed. This is done by taking grease from the tail gland and rubbing it over the plumage.

Feathers also serve as a coat that insulates the bird from overheating or undercooling. This coat must be carefully smoothed out.

Most important of all are the large wing quills and the tail feathers. The fibres making up the feather vanes must be re-hooked. An inefficient flyer would soon fall behind in the task of provisioning the fast-growing young.

PHOTOGRAPH LAWRENCE C. SHAFFER

The parent gulls show a curiously ambivalent attitude. They cannot resist the demands for food; but, at the same time, try to avoid being pecked-at and overrun.

PHOTOGRAPH LAWRENCE C. SHAFFER

For ever demanding food, the ravenous young give their parents no peace. When a parent returns from the feeding grounds with fresh supplies, the young birds scarcely leave time for the food to be regurgitated, but grab it from the parent's throat.

This is when the young have to apply the language of reassurance. Like the females during pair formation, they adopt hunched postures which signal: 'We mean no harm'.

Gradually the young gulls become more able to fend for themselves. They preen and stretch and fluff their wings—going through all the movements of their parents. Ever inquisitive, they forage about, pecking at everything they come across, finding out what is edible and what is not.

The territorial boundaries are still all-important. Young birds that wander outside are viciously pecked and chased away by neighbouring adults.

But the young will turn round and attack the same neighbours when those neighbours trespass on *their* territory. Although much stronger, the trespassing adults always run away. A profound respect for other people's homes is deeply ingrained in every inhabitant of the gullery.

The season is nearly over and the young birds are trying desperately to fly. In July, the whole colony is a whirl of beating wings as thousands of young gulls prance about making frantic efforts to get airborne—bouncing up and down just as though they were on pogo-sticks. Their wing movements improve gradually, not because the young *learn* to fly, but because their wings and flying muscles are growing stronger. One day the flying machinery reaches the stage when these flapping movements have success, and the young find themselves airborne.

Although birds do not have to learn the movements of flight as we have to learn to swim, the further refinement of their technique does require practice. Strong updraughts may carry inexperienced young high into the air. Unexpected turbulence causes them great trouble; and even in relatively quiet weather, a bird often lands downwind in a flurry of wings and legs.

It will be some time before the young gulls can achieve the effortless mastery of the air displayed by the tough and experienced adults. But they improve daily. Soon they can fly well enough to leave the nesting grounds and drift out with their parents to the distant shore. As they extend their range, they acquire invaluable experience. Full efficiency is achieved in three to four years—the time it takes them to reach full maturity.

Long before winter comes, the herring gulls will scatter along the coast; the lesser blackbacks fly south to Africa. And with them will fly their chocolate-coloured young—still beset by hostile neighbours, but now confident in defence.

Autumn settles over the dunes and the ruined castle beyond. Except for a few herring gulls, the reserve is silent and deserted.

And so it will remain. Until another spring, when the crying gulls return and life in 'Bird City' starts all over again.

Acknowledgements

Our thanks are due to the Lake District and Lancashire Naturalists' Trusts for permission to work on the Walney Island Nature Reserve; in particular to the Reserve's Secretary, J. Mitchell. We gratefully remember the many hours spent with the Warden of the Reserve, Walter Shepherd, and the invaluable help he gave us on numerous occasions. Lawrence C. Shaffer kindly allowed us to use the photographs on pp. 17, 21, 68, 71 and 72.

Further reading

For a general description of the life-history of the (closely related) herring gull, see
N. Tinbergen (1953) *The herring gull's world*. Collins, London.

For those interested in other aspects of bird biology we recommend
D. Lack (1968) *Ecological adaptations for breeding in birds*. Methuen, London.

A general treatment of the many types of social organization in animals is given in
N. Tinbergen (1953) *Social behaviour in animals*. Methuen, London.

For new information about signalling language in a variety of animal species, see
D. Burkhardt, W. Schleidt, and H. Altner (1967) *Signals in the animal world*. Allen and Unwin, London.

Photoset and printed in Great Britain by
BAS Printers Limited, Wallop, Hampshire